Forever Grateful To You:

Made in the USA
Columbia, SC
28 November 2021

49903510R00061